The Standard®
AVIONICS LOG

ASA-SA-V2

THE STANDARD® AVIONICS LOG
SA-V2

©1993 Aviation Supplies & Academics, Inc.

THE STANDARD® is a registered trademark of
Aviation Supplies & Academics, Inc.

ASA-SA-V2
ISBN 978-1-61954-183-2

Printed in the United States of America

[19] 24

Aviation Supplies & Academics, Inc.
7005 132nd Place SE
Newcastle, Washington 98059 USA
asa2fly.com | asa@asa2fly.com

TRANSPORTATION   USD $9.95

ISBN 978-1-61954-183-2

9 781619 541832

Registration Number _____

Logbook Number _____

From _____

To _____

Aircraft Make _____

Model _____

Serial Number _____

Date of Manufacture _____

Avionics Equipment

ITEM	MAKE	MODEL	PART NUMBER	SERIAL NUMBER	DATE		LOCATION STATION #	BATTERY DATES	
					INSTALLED	REMOVED		INSTALLED	REPLACE

Avionics Equipment

| ITEM | MAKE | MODEL | PART NUMBER | SERIAL NUMBER | DATE | | LOCATION STATION # | BATTERY DATES | |
					INSTALLED	REMOVED		INSTALLED	REPLACE

VOR Accuracy

MAKE _____ MODEL _____ SERIAL NUMBER _____ PART NUMBER _____

DATE	VOR #	LOCATION	TACH TIME	FREQUENCY	TYPE OF CHECK		BEARING ERROR		CHECKED BY	REMARKS
					GROUND	FLIGHT	+	-		

VOR Accuracy

MAKE _____ MODEL _____ SERIAL NUMBER _____ PART NUMBER _____

DATE	VOR #	LOCATION	TACH TIME	FREQUENCY	TYPE OF CHECK		BEARING ERROR		CHECKED BY	REMARKS
					GROUND	FLIGHT	+	-		

VOR Accuracy

MAKE _____ MODEL _____ SERIAL NUMBER _____ PART NUMBER _____

DATE	VOR#	LOCATION	TACH TIME	FREQUENCY	TYPE OF CHECK		BEARING ERROR		CHECKED BY	REMARKS
					GROUND	FLIGHT	+	-		

VOR Accuracy

MAKE _____ MODEL _____ SERIAL NUMBER _____ PART NUMBER _____

DATE	VOR#	LOCATION	TACH TIME	FREQUENCY	TYPE OF CHECK		BEARING ERROR		CHECKED BY	REMARKS
					GROUND	FLIGHT	+	-		

VOR Accuracy

MAKE _____ MODEL _____ SERIAL NUMBER _____ PART NUMBER _____

DATE	VOR#	LOCATION	TACH TIME	FREQUENCY	TYPE OF CHECK		BEARING ERROR		CHECKED BY	REMARKS
					GROUND	FLIGHT	+	-		

VOR Accuracy

MAKE _____ MODEL _____ SERIAL NUMBER _____ PART NUMBER _____

DATE	VOR#	LOCATION	TACH TIME	FREQUENCY	TYPE OF CHECK		BEARING ERROR		CHECKED BY	REMARKS
					GROUND	FLIGHT	+	-		

VOR Accuracy

MAKE _____ MODEL _____ SERIAL NUMBER _____ PART NUMBER _____

DATE	VOR#	LOCATION	TACH TIME	FREQUENCY	TYPE OF CHECK		BEARING ERROR		CHECKED BY	REMARKS
					GROUND	FLIGHT	+	-		

VOR Accuracy

MAKE _____ MODEL _____ SERIAL NUMBER _____ PART NUMBER _____

DATE	VOR #	LOCATION	TACH TIME	FREQUENCY	TYPE OF CHECK		BEARING ERROR		CHECKED BY	REMARKS
					GROUND	FLIGHT	+	-		

VOR Accuracy

MAKE _____ MODEL _____ SERIAL NUMBER _____ PART NUMBER _____

DATE	VOR #	LOCATION	TACH TIME	FREQUENCY	TYPE OF CHECK		BEARING ERROR		CHECKED BY	REMARKS
					GROUND	FLIGHT	+	-		

VOR Accuracy

MAKE _____ MODEL _____ SERIAL NUMBER _____ PART NUMBER _____

DATE	VOR #	LOCATION	TACH TIME	FREQUENCY	TYPE OF CHECK		BEARING ERROR		CHECKED BY	REMARKS
					GROUND	FLIGHT	+	-		

VOR Accuracy

MAKE _____ MODEL _____ SERIAL NUMBER _____ PART NUMBER _____

DATE	VOR #	LOCATION	TACH TIME	FREQUENCY	TYPE OF CHECK		BEARING ERROR		CHECKED BY	REMARKS
					GROUND	FLIGHT	+	−		

VOR Accuracy

MAKE _____ MODEL _____ SERIAL NUMBER _____ PART NUMBER _____

DATE	VOR #	LOCATION	TACH TIME	FREQUENCY	TYPE OF CHECK		BEARING ERROR		CHECKED BY	REMARKS
					GROUND	FLIGHT	+	-		

GPS Database

MAKE _____ MODEL _____ SERIAL NUMBER _____ PART NUMBER _____

DATABASE UPDATED	DATABASE EXPIRES	UPDATED BY	REMARKS

GPS Database

MAKE _____ MODEL _____ SERIAL NUMBER _____ PART NUMBER _____

DATABASE UPDATED	DATABASE EXPIRES	UPDATED BY	REMARKS

GPS Database

MAKE _____ MODEL _____ SERIAL NUMBER _____ PART NUMBER _____

DATABASE UPDATED	DATABASE EXPIRES	UPDATED BY	REMARKS

GPS Database

MAKE _____ MODEL _____ SERIAL NUMBER _____ PART NUMBER _____

DATABASE UPDATED	DATABASE EXPIRES	UPDATED BY	REMARKS

GPS Database

MAKE _____ MODEL _____ SERIAL NUMBER _____ PART NUMBER _____

DATABASE UPDATED	DATABASE EXPIRES	UPDATED BY	REMARKS

GPS Database

MAKE _____ MODEL _____ SERIAL NUMBER _____ PART NUMBER _____

DATABASE UPDATED	DATABASE EXPIRES	UPDATED BY	REMARKS

GPS Database

MAKE _____ MODEL _____ SERIAL NUMBER _____ PART NUMBER _____

DATABASE UPDATED	DATABASE EXPIRES	UPDATED BY	REMARKS

GPS Database

MAKE _____ MODEL _____ SERIAL NUMBER _____ PART NUMBER _____

DATABASE UPDATED	DATABASE EXPIRES	UPDATED BY	REMARKS

GPS Database

MAKE _____ MODEL _____ SERIAL NUMBER _____ PART NUMBER _____

DATABASE UPDATED	DATABASE EXPIRES	UPDATED BY	REMARKS

GPS Database

MAKE _____ MODEL _____ SERIAL NUMBER _____ PART NUMBER _____

DATABASE UPDATED	DATABASE EXPIRES	UPDATED BY	REMARKS

GPS Database

MAKE _____ MODEL _____ SERIAL NUMBER _____ PART NUMBER _____

DATABASE UPDATED	DATABASE EXPIRES	UPDATED BY	REMARKS

GPS Database

MAKE _____ MODEL _____ SERIAL NUMBER _____ PART NUMBER _____

DATABASE UPDATED	DATABASE EXPIRES	UPDATED BY	REMARKS

Transponder

MAKE _____ MODEL _____ SERIAL NUMBER _____ PART NUMBER _____
CHECK MODES: A _____ C _____ S _____   REQUIRED INSPECTION FREQUENCY _____

DATE	TYPE OF CHECK	NEXT CHECK DUE	REPAIR SHOP AND NUMBER	SIGNATURE

Transponder

MAKE _____ MODEL _____ SERIAL NUMBER _____ PART NUMBER _____

CHECK MODES: A _____ C _____ S _____ REQUIRED INSPECTION FREQUENCY _____

DATE	TYPE OF CHECK	NEXT CHECK DUE	REPAIR SHOP AND NUMBER	SIGNATURE

Emergency Locator Transmitter

DATE	TYPE OF UNIT	PART # / SERIAL #	TRANSMIT TEST	CHECK CONDUCTED BY	BATTERY DUE DATE

Emergency Locator Transmitter

DATE	TYPE OF UNIT	PART # / SERIAL #	TRANSMIT TEST	CHECK CONDUCTED BY	BATTERY DUE DATE

Airworthiness Directive Compliance

A.D. NUMBER	SUBJECT	COMPLIANCE Date, Time or Cycles	METHOD OF COMPLIANCE	TYPE A.D. 1	TYPE A.D. R	NEXT COMPLIANCE DUE	AUTHORIZED SIGNATURE / CERTIFICATE NUMBER

1 = One Time
R = Recurring

Airworthiness Directive Compliance

A.D. NUMBER	SUBJECT	COMPLIANCE Date, Time or Cycles	METHOD OF COMPLIANCE	TYPE A.D. 1	TYPE A.D. R	NEXT COMPLIANCE DUE	AUTHORIZED SIGNATURE / CERTIFICATE NUMBER

1 = One Time
R = Recurring

Airworthiness Directive Compliance

A.D. NUMBER	SUBJECT	COMPLIANCE Date, Time or Cycles	METHOD OF COMPLIANCE	TYPE A.D. 1	TYPE A.D. R	NEXT COMPLIANCE DUE	AUTHORIZED SIGNATURE / CERTIFICATE NUMBER

1 = One Time
R = Recurring

Airworthiness Directive Compliance

A.D. NUMBER	SUBJECT	COMPLIANCE Date, Time or Cycles	METHOD OF COMPLIANCE	TYPE A.D. 1	TYPE A.D. R	NEXT COMPLIANCE DUE	AUTHORIZED SIGNATURE / CERTIFICATE NUMBER

1 = One Time
R = Recurring

Factory Service Bulletins

MANUFACTURER	BULLETIN NUMBER	BULLETIN DATE	REVISION #	SUBJECT	COMPLIANCE Date, Time or Cycles	NEXT COMPLIANCE Date, Time or Cycles

Factory Service Bulletins

MANUFACTURER	BULLETIN NUMBER	BULLETIN DATE	REVISION #	SUBJECT	COMPLIANCE Date, Time or Cycles	NEXT COMPLIANCE Date, Time or Cycles

Factory Service Bulletins

MANUFACTURER	BULLETIN NUMBER	BULLETIN DATE	REVISION #	SUBJECT	COMPLIANCE Date, Time or Cycles	NEXT COMPLIANCE Date, Time or Cycles

Factory Service Bulletins

MANUFACTURER	BULLETIN NUMBER	BULLETIN DATE	REVISION #	SUBJECT	COMPLIANCE Date, Time or Cycles	NEXT COMPLIANCE Date, Time or Cycles

Maintenance Record

DATE	ITEM	SERIAL #	DISCREPANCY	CORRECTIVE ACTION OR MODIFICATION	REPAIR SHOP AND NUMBER

Maintenance Record

DATE	ITEM	SERIAL #	DISCREPANCY	CORRECTIVE ACTION OR MODIFICATION	REPAIR SHOP AND NUMBER

Maintenance Record

DATE	ITEM	SERIAL #	DISCREPANCY	CORRECTIVE ACTION OR MODIFICATION	REPAIR SHOP AND NUMBER

Maintenance Record

DATE	ITEM	SERIAL #	DISCREPANCY	CORRECTIVE ACTION OR MODIFICATION	REPAIR SHOP AND NUMBER

Maintenance Record

DATE	ITEM	SERIAL #	DISCREPANCY	CORRECTIVE ACTION OR MODIFICATION	REPAIR SHOP AND NUMBER

Maintenance Record

DATE	ITEM	SERIAL #	DISCREPANCY	CORRECTIVE ACTION OR MODIFICATION	REPAIR SHOP AND NUMBER

Maintenance Record

DATE	ITEM	SERIAL #	DISCREPANCY	CORRECTIVE ACTION OR MODIFICATION	REPAIR SHOP AND NUMBER

Maintenance Record

DATE	ITEM	SERIAL #	DISCREPANCY	CORRECTIVE ACTION OR MODIFICATION	REPAIR SHOP AND NUMBER

Maintenance Record

DATE	ITEM	SERIAL #	DISCREPANCY	CORRECTIVE ACTION OR MODIFICATION	REPAIR SHOP AND NUMBER

Maintenance Record

DATE	ITEM	SERIAL #	DISCREPANCY	CORRECTIVE ACTION OR MODIFICATION	REPAIR SHOP AND NUMBER

Maintenance Record

DATE	ITEM	SERIAL #	DISCREPANCY	CORRECTIVE ACTION OR MODIFICATION	REPAIR SHOP AND NUMBER

Maintenance Record

DATE	ITEM	SERIAL #	DISCREPANCY	CORRECTIVE ACTION OR MODIFICATION	REPAIR SHOP AND NUMBER

Maintenance Record

DATE	ITEM	SERIAL #	DISCREPANCY	CORRECTIVE ACTION OR MODIFICATION	REPAIR SHOP AND NUMBER

Maintenance Record

DATE	ITEM	SERIAL #	DISCREPANCY	CORRECTIVE ACTION OR MODIFICATION	REPAIR SHOP AND NUMBER

Maintenance Record

DATE	ITEM	SERIAL #	DISCREPANCY	CORRECTIVE ACTION OR MODIFICATION	REPAIR SHOP AND NUMBER

Maintenance Record

DATE	ITEM	SERIAL #	DISCREPANCY	CORRECTIVE ACTION OR MODIFICATION	REPAIR SHOP AND NUMBER

Maintenance Record

DATE	ITEM	SERIAL #	DISCREPANCY	CORRECTIVE ACTION OR MODIFICATION	REPAIR SHOP AND NUMBER

Maintenance Record

DATE	ITEM	SERIAL #	DISCREPANCY	CORRECTIVE ACTION OR MODIFICATION	REPAIR SHOP AND NUMBER

Maintenance Record

DATE	ITEM	SERIAL #	DISCREPANCY	CORRECTIVE ACTION OR MODIFICATION	REPAIR SHOP AND NUMBER

Maintenance Record

DATE	ITEM	SERIAL #	DISCREPANCY	CORRECTIVE ACTION OR MODIFICATION	REPAIR SHOP AND NUMBER

Maintenance Record

DATE	ITEM	SERIAL #	DISCREPANCY	CORRECTIVE ACTION OR MODIFICATION	REPAIR SHOP AND NUMBER

Maintenance Record

DATE	ITEM	SERIAL #	DISCREPANCY	CORRECTIVE ACTION OR MODIFICATION	REPAIR SHOP AND NUMBER